Content

1)whats is iss_____page 1

2)space shuttle_____page 4

3) Soyuz_____page 6

4)how to be astronaut__page 8

5) training_____page 10

6)launch_____ page 12

7)working_____page 19

8)sleeping_____page 21

9)dining area _____page 23

10)copula_____page 26

11)iss gym_____page 27

12)health_____page 29

13)accidents in space____page 36

14)free time _____page 38

15)landing _____page 40

Shortcuts

NASA=national aeronautic space agency

ESA= European space agency

JAXA = japan space agency

Iss=international space station

What is the iss?

The iss is an artificial satellite which consist of many parts first part was send in 1998 November by Russia then other nations and Russia continued building it the iss until 2011

  the first man in the iss was Yuri Gagarin the size of the iss is the size of football field it weighs in the earth around 1 million pound iss orbit the earth at the speed of 28,000km/h (17,500mile/h)which means it completes one rotation around the in just 90 minutes Which means there is 16 sun rise and 16 sun set in a day its found 400km above the earth it works by solar power the iss consist of Utility mogul,bathroom,gym and it have 2 big arms and 10 laboratory mogul Russia and us has 4 each while one for Europe and other for japan its used to to make experiment in zero gravity environment and make research on human can we live in zero G or not but know they know yes we can and experiment in bacteria can they live in harsh environment there are 6 astronaut and cosmonaut and tourist living in the iss together for 6 month the different between cosmonaut and astronaut is that cosmonaut are Russian while astronaut are from NASA,

JAXA,ESA

Space shuttle

is a machine or special type of jet that take cosmonaut and astronaut to the iss it can take up to 7 crews at a time it lunches from USA it consists Of 3 parts the main which is the only part that stays in orbit the wait on which look like plane the second is the external thank which holds of fuel and the 3rd are the side rockets which lunches for the first 2 minutes to increase the power

Fuel tank 2nd part

3rd part

Main part

Soyuz

IS Russian spacecraft which is different from space shuttle but both of them have same function The power of the iss is 30 million horse power Soyuz can hold 3 crew it lunches from Russia it have three parts the first is place where crew sit when it lunch and land that is the only part that return to earth the other will remain at the atmosphere and it will burn 2nd part is the place where the crews sleep eat and go to bathroom while the 3rd is the place where engines and oxygen are found

How to be astronaut

How to be astronaut first and main thing you must be good in physics at your high school and you must have high GDP when you graduate you must attend university and take physic,math or engineering course when you finish your course you must have 3 year experience of your course or 3000 hours of jet flying record before you apply to be astronaut after that you will apply for being accepted you must talk English very well and your eye sight must be 20/20 And your height must be less than 190cm and above 160 cm

TRANING

After being accepted for being astronaut NASA or other space agencies will not consider you as a full astronaut until you complete the training will take 2 years your training the first training is to keep your mind concentrated under stress they will let you perform parachute jumping while solving math problems then you will drive a space shuttle simulator after that you will perform space walk under water at hydro zero gravity , you will experience illusion flight laboratory at zero g plane that last no more than 30 second and they will fly jets and many other

launch

before a day or more from the lunching day the rocket will be placed at the lunching pad after the refueling the rocket will start breathing it will be covered by liquid oxygen naturally After the crew finish there training they will wait until the lunching day no one will not know the exact time but the mission control center will estimate the time and the crew will be told exactly 8 hour before lunching then they will

Tell you its time to go and you will start wearing there space suits behind a glass so there family and other people could see us the will wave for you and take picture for you that's because they are proud at that time you will have mixed feelings of joy and nervous and sudden question will come in your head why why am I hear am is it must to go there must I take all this risk and still you can refuse to go. then it will be time for going you will leave your autograph at the door then

you are good to go   you will
go by a bus to the lunching pad
you go up to the ninth floor by
elevator and it will be point of
no return with only space ahead
of you after you go inside the
shuttle you go some steps and
there you will find a red carpet
like hotels but you must crawl
in there then you will sit in your
positions for 2 hours until
lunching they will check
everything again then count
down will start and lift off the
journey which will take 9
minutes to space and 2 days to

iss will start when you reach the atmosphere you will feel like four people like you are siting in your chest after that zero gravity will kick on and every stationary object will start floating with the dust and you feel the relief until you reach there you are must to sleep more eat less until you start orbit you don't have water for other natural needs you don't need to be patient you have pampers for that modern technology allows using them so why not for the first 24 hours you will hauted by the feeling of someone big and strong taking your legs up to your ears its like you have been head over heels

Zero g is really hard you have long
time to experience it in training
communicating with the land is
hard its always interrupted by
Chinese song or Indian movie
scene then they will dock with the
iss its hard because the land and
the crew inside the iss are under
risk of the shuttle crew if they
make mistake all will be in danger
Slowly they will dock the hatch
opening is the most exiting time
You will meet people and they will
ask the new crew news from earth
that they cant ask in phone or

email and the crew will eat and drink hot food and drinks two days without hot food and drink in the cold ship is really hard after that they will refresh they cant take bath there but they will shave wash there face and change clothes while the shuttle been small the station seems big even huge.

Working

Astronaut perform many experiments up there some might be biological some physics and others like what will happen to human in zero g there are many laboratories there its nice working in zero g because you don't need to stand or sit you will be floating you will not be tiered when you are working in computer or performing researches you need your leg for fixation parpeses only but the change in function of legs in space leads for the change of them too in space something will change in your system

<u>Sleeping</u>

Astronaut get enough sleep there the only difference is that they sleep standing there you know there is no gravity so you don't feel either you are standing or leaning but every human want to feel something hard in his back so they tie their back to the wall by some special robe they have there own room with

sleeping bag ,personal laptops and their photos its comfortable to sleep there than on earth

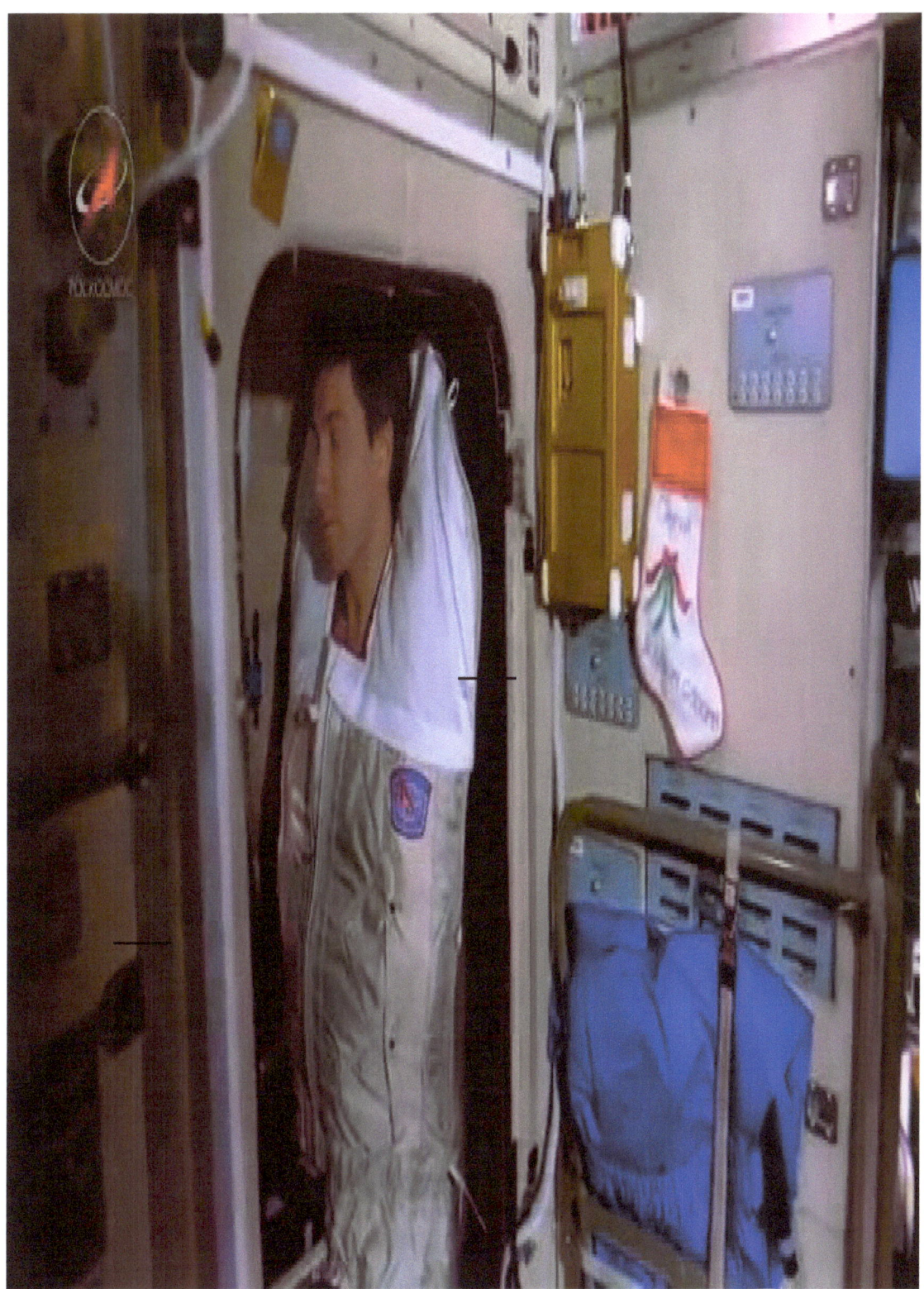

Dinning area or kitchen

Astronaut have any type of food they want egg,meat , vegetable and etc since its possible to delivery anything to the iss astronaut meet at the evening and morning at the dinning area for food so they talk about what they did trough out the day if American astronaut don't like there food which is in a blue box they will look for the Russian which is in the red 90% of the crew are fun of the Russian cottage chess so it works as a exchange currency there if a Russian likes something

from the us,European or Japanese
Food he could exchange it
astronauts receives a special
delivery every 6 weeks like
food,cloth and new experiments
they receive some from family and
friend while other from the space
agency so they can order there
family special food the most
expensive thing in orbit is water

Copula

Copula is a window found in the iss that shows the earth 360 degree its used for entertainment,for seeing and catching the Soyuz and taking picture of countries

Gym

Exercise is not must at the iss but if your willing to return to earth healthy you must exercise or your bone and muscle become weak Astronaut perform many exercise they must perform at least 2 hours per day they perform cycling ,walking machine and weight lifting exercise is way more harder in space than earth

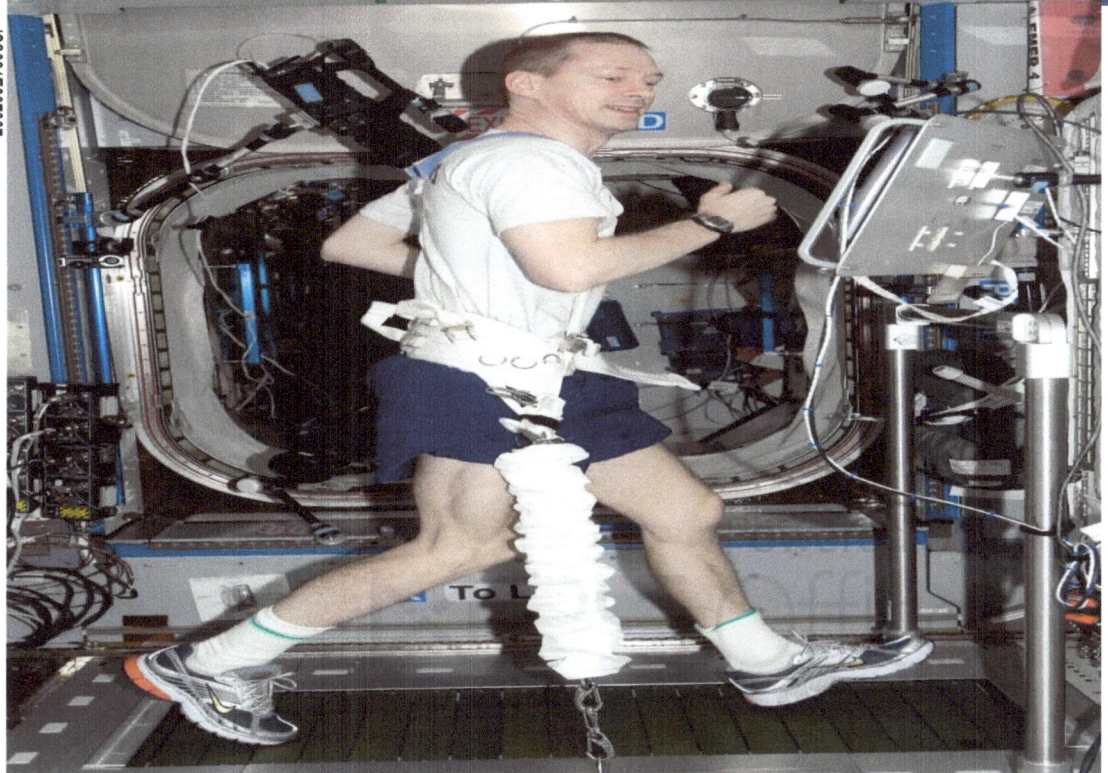

health

A human physiology study that will keep the crew busy many days of their mission is Ocular Health. Its full name is Prospective Observational Study of Ocular Health in ISS Crews. The Prospective Observational Study of Ocular Health in ISS Crews (Ocular Health) protocol aims to systematically gather physiological data to characterize the Risk of Microgravity-Induced Visual Impairment/Intracranial Pressure on crewmembers assigned to a 6 month ISS increment," the NASA experiment overview said.

It is known that some (not all) astronauts in orbit experience changes in visual acuity (visual clarity) and intraocular pressure as a result of fluid shifts within the body as it is subjected to micro gravity. About 20% is astronauts flying to ISS have reported these kinds of changes. Test subjects will undergo pre-flight, flight and post-flight testing of their eyes using a variety of techniques.

Near and far visual acuity will be tested for each eye independently using a Snellen chart and Amsler grid testing. Also, intraocular and blood pressure measurements are part of the study as well as ocular ultrasounds to identify changes in globe morphology, fundoscopy to detect retinal changes, threshold visual field testing to assess central and peripheral vision changes, contrast sensitivity changes as a measure of visual function, and vascular compliance calculations. In-flight measures will be taken on Flight Days 10, 30, 60, 90, and 120 as well as 30 days prior to return.

Pre- and post-flight examinations include refraction testing, pupil reflex testing, extraocular muscle balance and function checks, CT and MRI scans and split lamp biochemistry and high resolution retinal photography.

"The purpose of this study is to collect evidence to characterize the risk and define the visual changes and central nervous system (CNS) changes observed during a six month exposure to micro gravity including post flight time course for recovery to baseline. This study will gather information that can be used to assess the risk of Micro gravity-Induced Visual Impairment / Intracranial Pressure (VIIP) and guide future research needs," the experiment overview notes. Mainly weightlessness gradually destroy the human body

Spacewalk

Spacewalk is one of the most dangers things in space astronaut perform spacewalk when there is problem in the iss and the earth control center cant fix it astronaut train for many month to perform space walk there must be more than one astronaut to perform spacewalk at time the others will help them put the space suit and close the hatch astronaut spends hours working outside they are tide to not get lost and always the one will check his mate if his fine

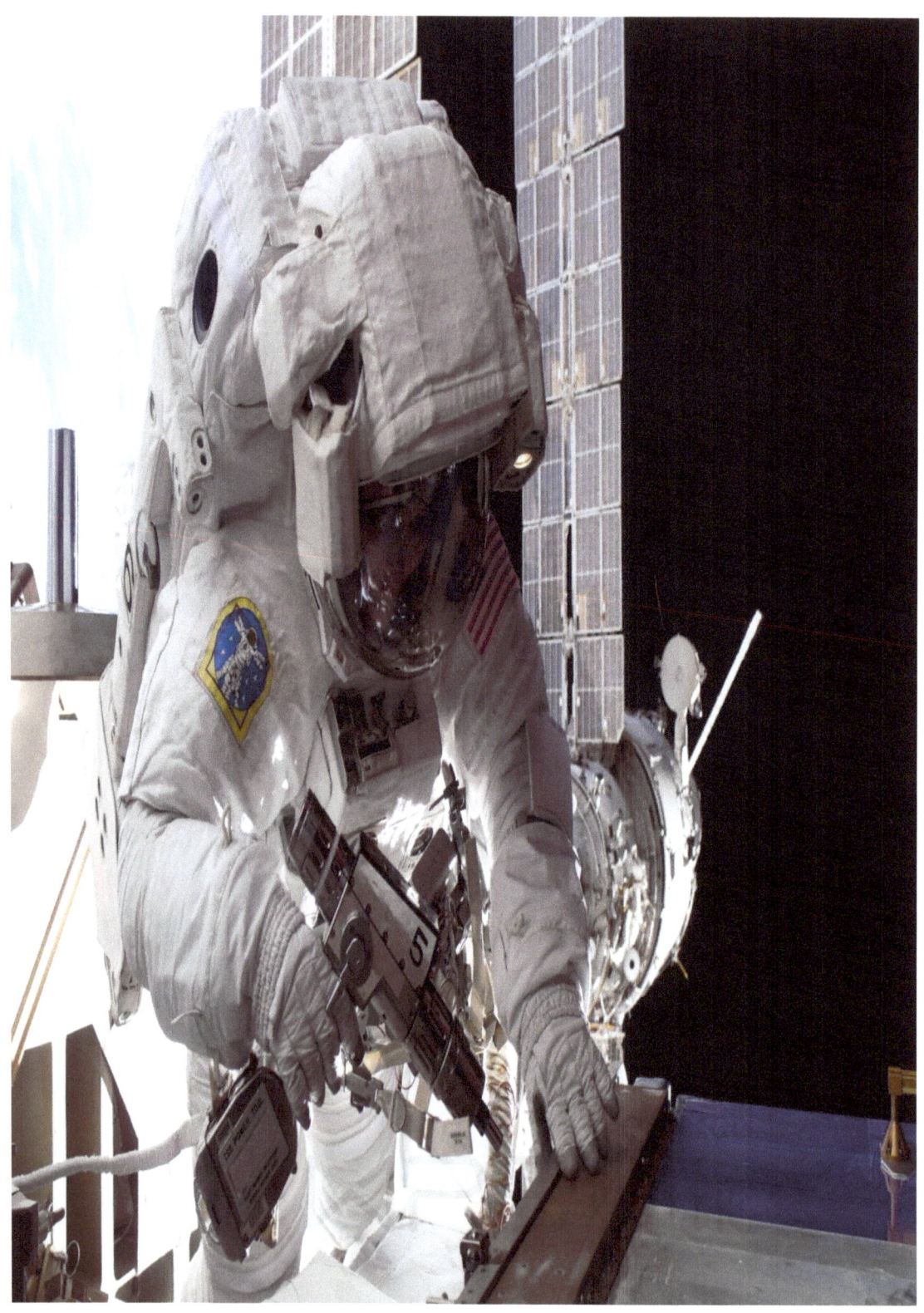

Accidents at space

Accidents at space are very
dangers which will lead to death
There are two shuttles in the iss
the are like life boats accidents in
space are like lunching accident's
example:challengers disaster ,fire ,
depression, re entry accidents
All are dangers but astronaut are
trained very well to pass this
accidents

Challengers disaster

Weekend,holiday and free time

There is free time in space too so have you every thought how do astronaut spend it in the mid of the month They cant go to barber to so they cut each others hair by using special machine which cleans all the hair to not get to machines And almost all Saturday is cleaning day astronaut clean the iss by them self because there is no one else to clean and fix things up there they fix bathroom by them self and when they finish everything they might play games like

football,chess etc
They spend time in cupola to
they play games there like
guessing where they are and many
more they can call their family and
friends too using email and phone
call they can use internet too
They cant go to barber to so they
cut each others hair by using
special machine which cleans all
the hair to not get to machines

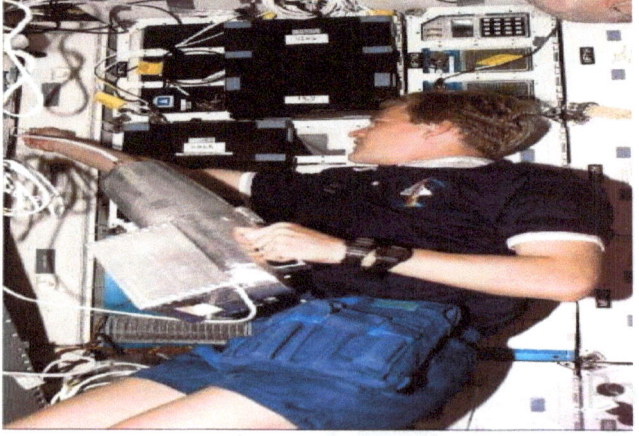

Landing

Landing is one of the dangerous and hard part of work as an astronaut you will say bye to your Friends up their and they will get in to the hatch and they will air lock astronauts will wear there space suits inside the shuttle then they will sit in there special chairs then they will undock then they must correct the ships altitude then the other two parts will undock and the control Mogul is the only part that will return to earth the ship must enter the atmosphere by the right position otherwise it will burn when it enter the atmosphere the down part of the part of the ship starts burning the crew can see it from the window the gravity will hit on everything will be hard even your hands because you adapted to zero gravity then there chairs will rise up there helmet will stick to the control board before undocking helicopters will start flying to find the ship faster they will make the landing place empty to make the landing smooth which is actually not smooth its like you are in small car and you crash with a large truck astronaut will train how to survive in any place until other people find them they might land in water,forest, desert or cold mountains Chance of landing on land is

1/3 the rest are water after landing they will come and open for you the hatch it will be hot in because its was burning astronauts could lose up-to 5 kg after they opened the hatch you will see your doctor that you has been talking with for six month they will hold you and put you in a chair you will feel bad feelings you cant see properly because you adapted to zero g they will give apple you be extremely happy everything will change in your life after going to space you will be special then they will take you to home after meeting everyone and breathing enough air of earth in one month you will start searching for stars and you will say am I ready for other flight

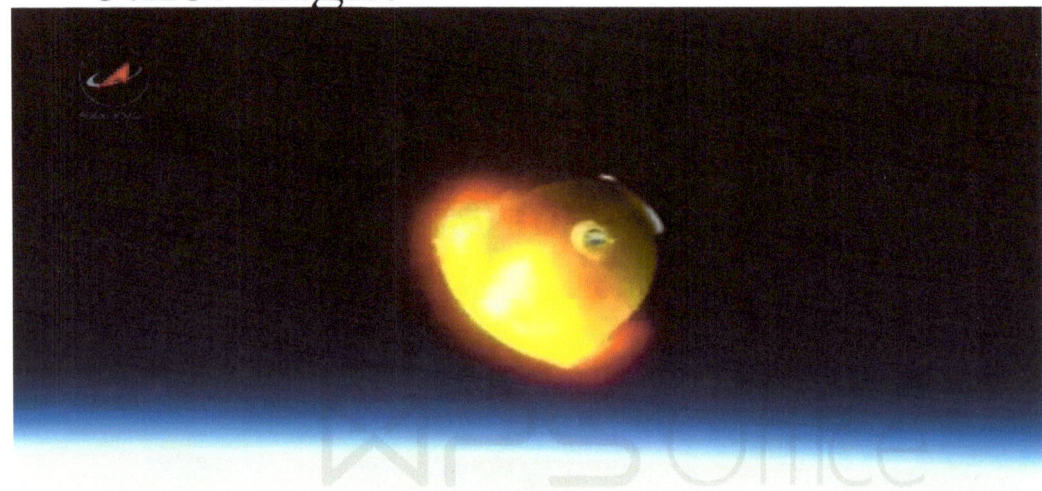

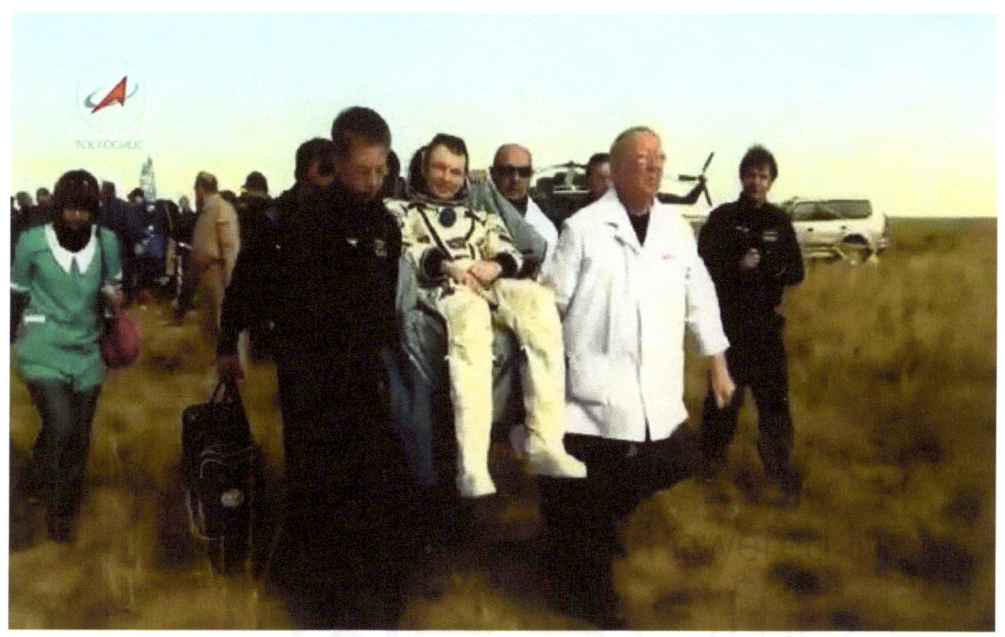

The end